610 Historical Portal

+ CC '81

Two cells meet—the mother's ovum and the father's sperm—becoming one life-bearing cell. And so begins the truly incredible story of the unborn baby. This first cell is so small that it could fit on the head of a pin. Yet, over the next nine months, it will grow, remarkably, into a new human being.

With soft and lovely watercolor drawings, Colette Portal evokes the beauty of this development and growth. Tiny arm buds become arms with grasping hands; a child's sleeping face emerges from a strange and wrinkled shape. Finally, the fully-developed baby leaves its watery world and is born into the outside world.

Step by step, the story of human reproduction unfolds dramatically in a simple presentation that offers wide appeal for all ages.

Colette Portal, a young French painter and illustrator of *The Honeybees* and *Life of a Queen*, lives in Burcy, France, with her husband, Jean-Michel Folon. While researching *The Beauty of Birth* she collaborated with two French physicians.

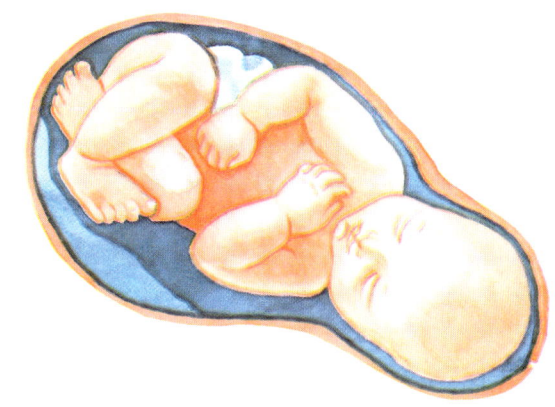

THE BEAUTY OF BIRTH

Colette Portal

Adapted from the French by Guy Daniels

ALFRED · A · KNOPF  NEW YORK

This is a Borzoi Book published by Alfred A. Knopf, Inc.
Copyright © 1971 by Colette Portal
All rights reserved under International and Pan-American Copyright Conventions.
Published in the United States by Alfred A. Knopf, Inc., New York,
and simultaneously in Canada by Random House of Canada Limited, Toronto.
Distributed by Random House, Inc., New York.
Trade Edition: ISBN: 0-394-82287-0
Library Edition: ISBN: 0-394-92287-5
Library of Congress Catalog Card Number: 70-155813
Manufactured in the United States of America. First Edition.

I wish to thank Jean-Michel Folon and the doctors, Pierre Simon and Alfred Tomatis, for the help they extended to me while I was writing this book.

Colette Portal

I am grateful to Dr. Robert Landesman for his assistance in verifying the medical information of the English translation.

Guy Daniels

INTRODUCTION

A baby is born of the love between a man and a woman. Like all love stories human life begins with a chance meeting. Two cells come together: one, the sperm from the father; and the other, the ovum, from the mother.

With his sex organ, called the penis, the man deposits his sex cells into the woman's sex organ, called the vagina. When one sperm and the ovum meet, they merge into a single cell: the human egg.

The following pages describe this meeting and the development of the egg into a human being.

 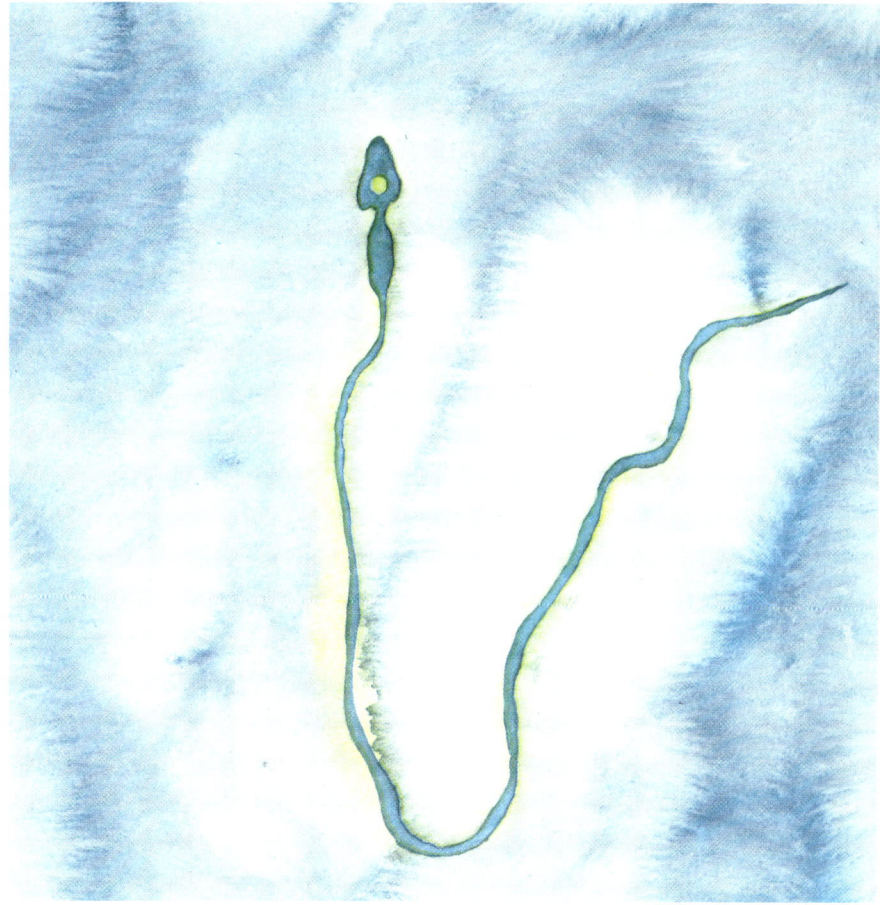

THE OVUM

The sex cell from the mother is called an ovum. It is round, and enclosed in a protective covering. The center of the ovum contains the cell's twenty-three chromosomes—tiny "packets" which will determine the particular characteristics of the future child.

The ovum is large and passive, contrasting to the minute and frenetic sperm from the male.

THE SPERM

The sex cell from the father is called a sperm. It has a head and tail, and is shaped like a tadpole. The sperm can swim very fast and is very tiny. In fact, it takes 85,000 sperm to equal the size of one ovum!

In the head of the mature sperm is the same number of chromosomes as in the nucleus of the mature ovum: twenty-three.

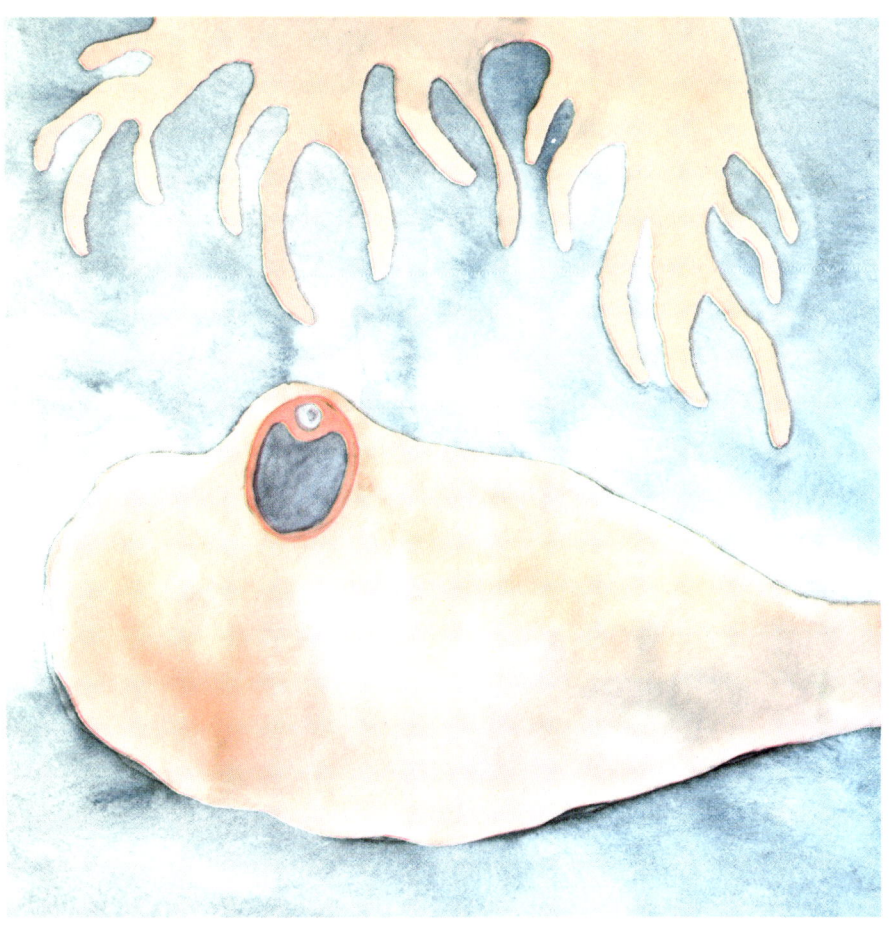

AN OVARY

The ovum develops from an immature egg cell. The egg cells are formed in two glands known as the ovaries. There is one ovary on either side of a woman's uterus, the cavity low in her abdomen in which the baby will eventually grow before birth.

Every month, one of these egg cells grows into an ovum. And at first it is sheltered in a little bag, or follicle, within the ovary.

OVULATION

But the bag soon bursts, like a ripe fruit spilling its seed. The young ovum escapes and begins its free life. This is called ovulation.

A FALLOPIAN TUBE

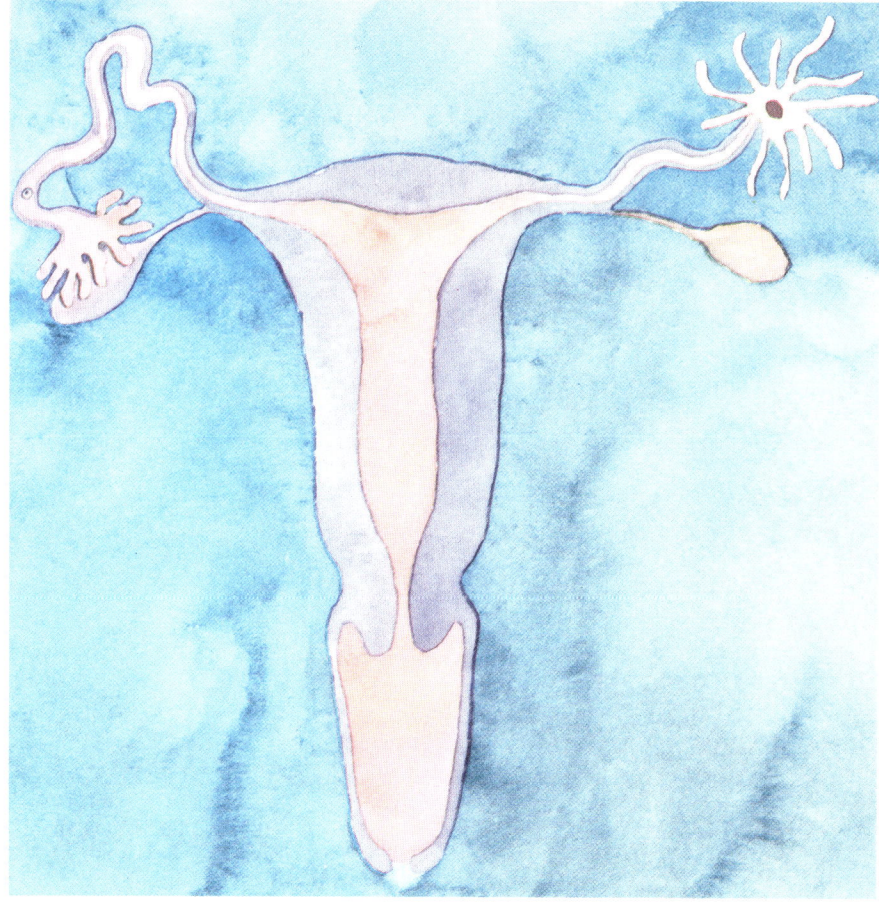

THE UTERUS

No sooner has it escaped, however, than the ovum falls into a kind of cup with fringed edges which forms one end of a tube called the Fallopian tube. The ovum then begins to move toward the other end of the Fallopian tube, which opens into the uterus.

There are two Fallopian tubes, one connecting each ovary to the uterus.

Once out of the ovary and into the Fallopian tube, the slow-moving ovum begins its journey down the Fallopian tube toward the uterus.

THE PATH THROUGH THE FALLOPIAN TUBE

The ovum is carried along its path by fluids. This path is bordered by cilia, which resemble little hairs.

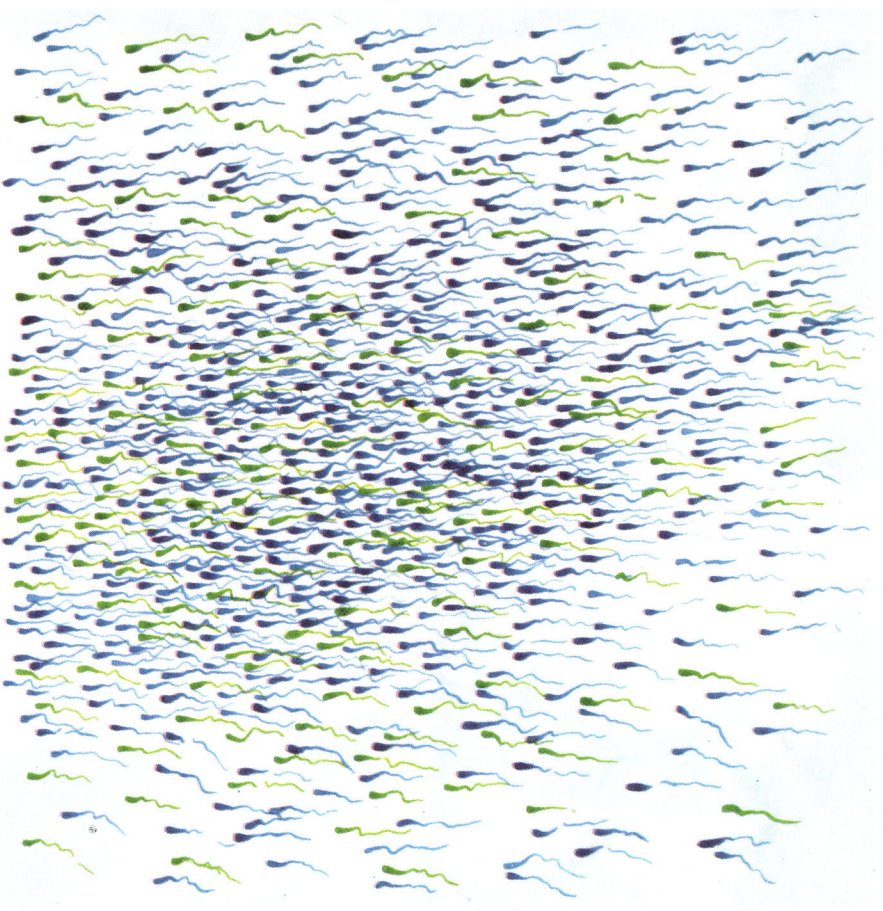

THE MOVEMENT OF THE SPERM

The very active sperm have a long way to travel before reaching the place of fertilization in the Fallopian tube. Sperm are produced in the testicles of the man, the two sex glands located at the base of the penis. Once deposited in the vagina, they begin swimming in all directions. But the movements of their heads and the lashing of their tails drive some of them up toward the uterus.

THE OBSTACLES

The sperm swim upstream against a downward current caused by the beating of cilia in the lining of the uterus. Although millions of sperm are deposited in the vagina, most of them go astray in the many folds and wrinkles of the vagina and the uterus. But a few thousand reach the place of fertilization in the Fallopian tube, having traveled a path bristling with obstacles.

THE MEETING

Finally, the sperm meet the ovum. Thousands of sperm cling to the ovum's protective covering.

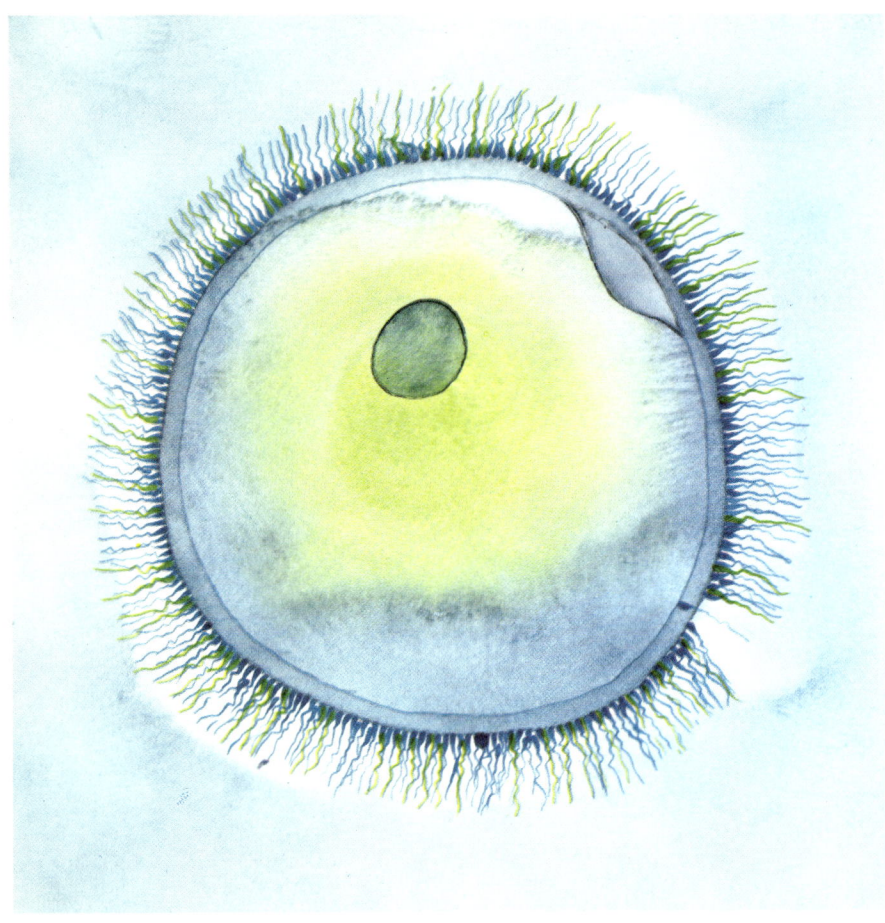 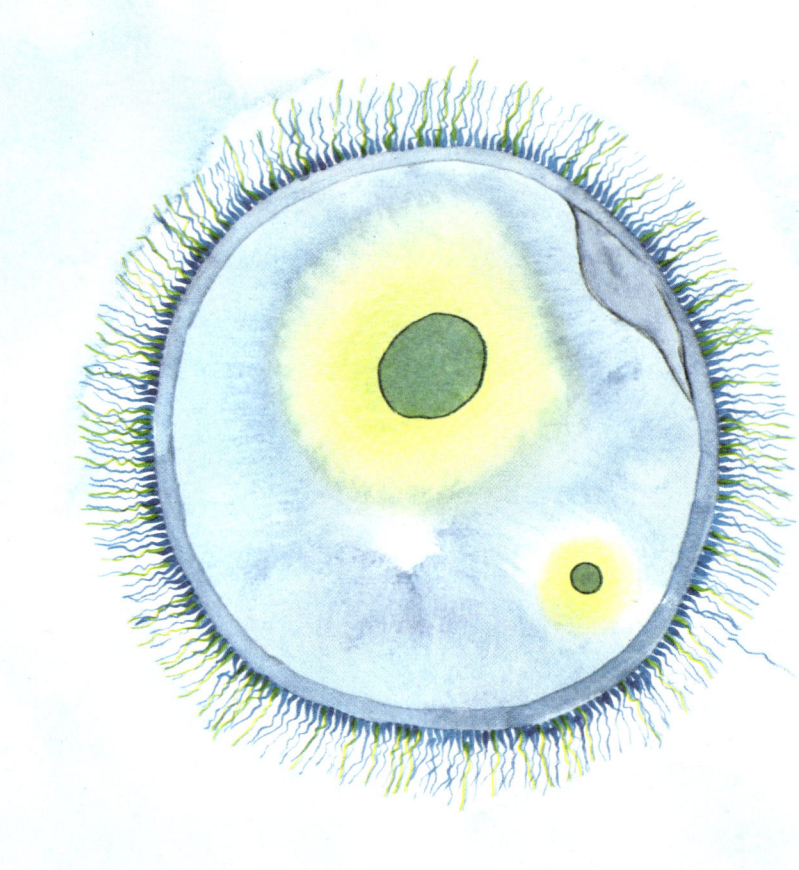

FERTILIZATION

Only one sperm manages to drive its head into the ovum, and it becomes the "winner"—the sperm which will pass on its twenty-three chromosomes to the new human being.

In the nuptial flight of bees, the queen chooses as her mate the male which can fly the highest. But in the case of sperm, it is sheer chance that determines which one will penetrate the ovum.

THE MALE NUCLEUS AND THE FEMALE NUCLEUS

As soon as the head of the first sperm enters the ovum, its tail drops off. At that very instant, the ovum seals itself off from other sperm.

During the next few minutes the head of the sperm enlarges, forming a nucleus which approaches the nucleus of the ovum.

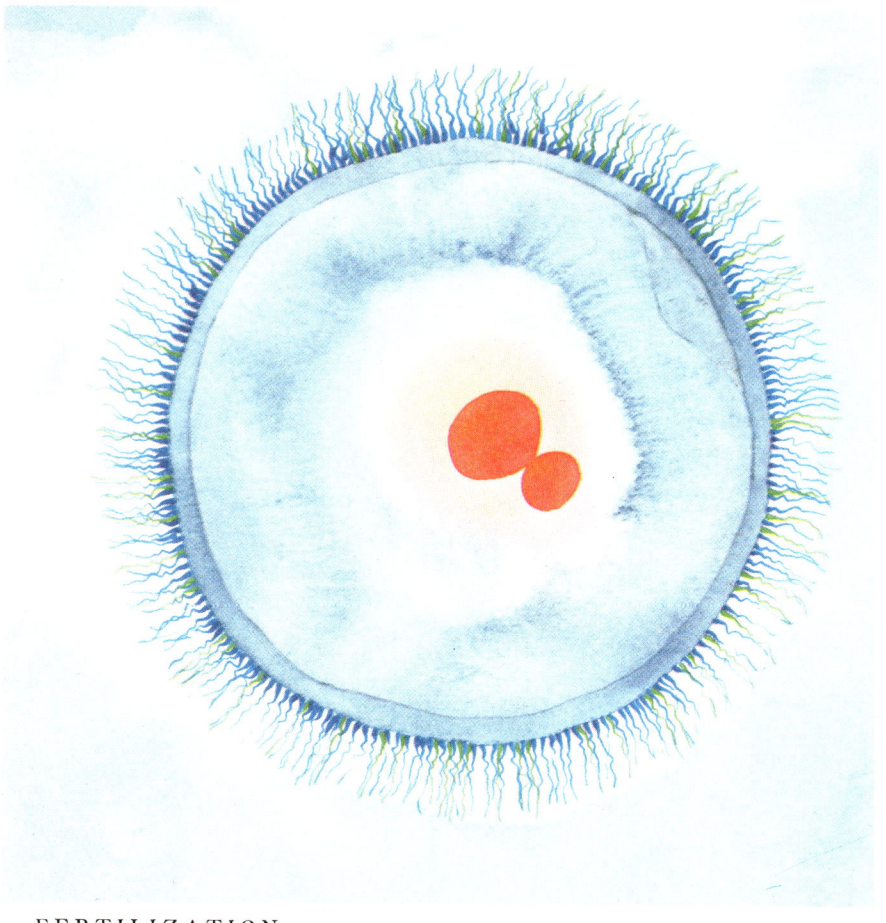

FERTILIZATION

When the two nuclei come together they merge to become a single nucleus. This new nucleus now has forty-six chromosomes—twenty-three from the ovum and twenty-three from the sperm.

Now the ovum is mature, it has become an egg—a life-bearing cell. Fertilization has taken place.

THE EGG

Three hours after fertilization, the egg begins to divide.

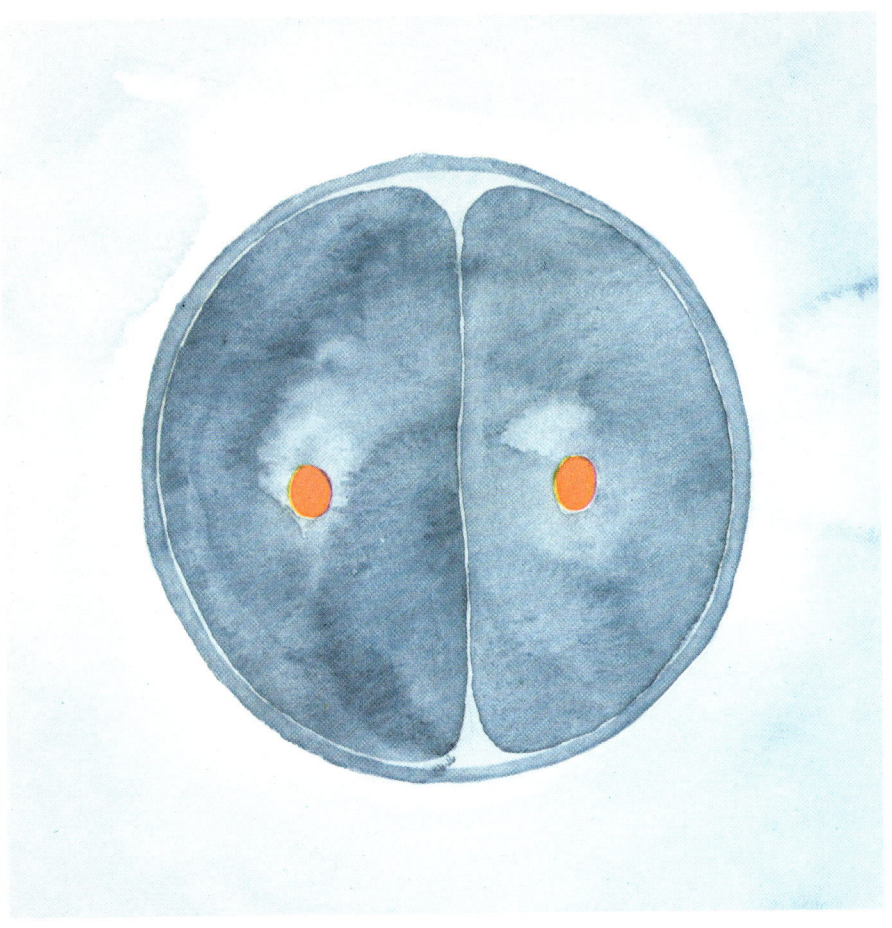

TWO CELLS

Thirty hours after fertilization, the egg divides down the middle into two identical cells. Each cell has a nucleus.

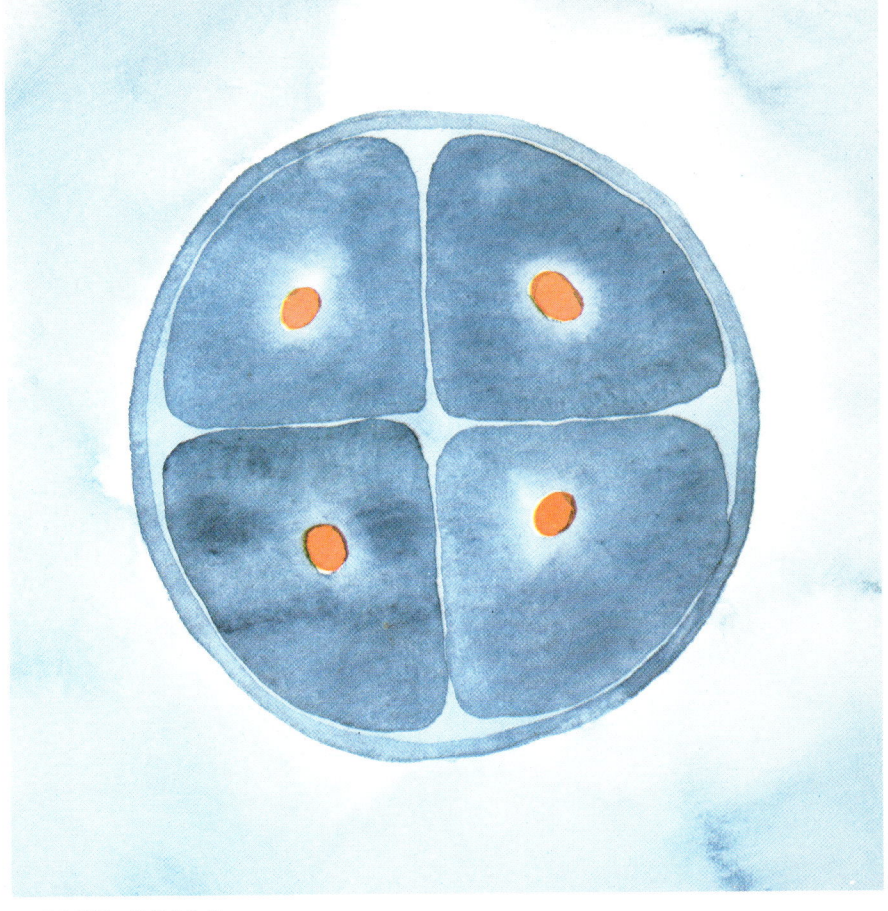

FOUR CELLS

After a period of rest, the two cells divide again to become four.

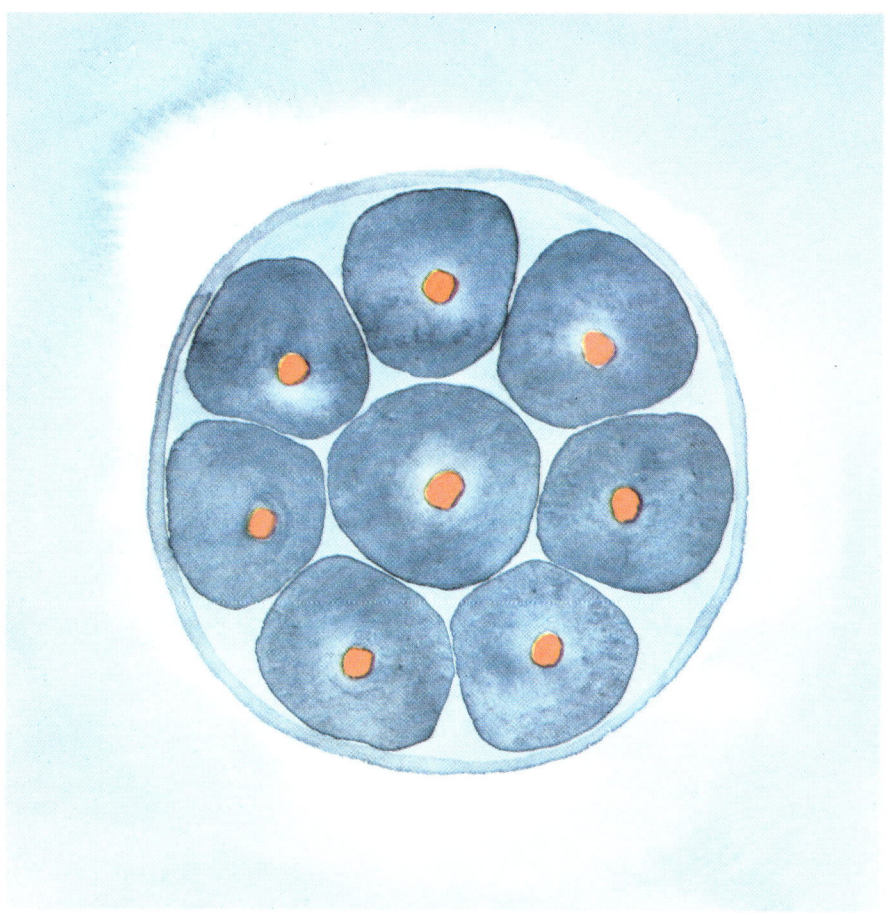

EIGHT CELLS

Another pause, and the egg has eight cells.

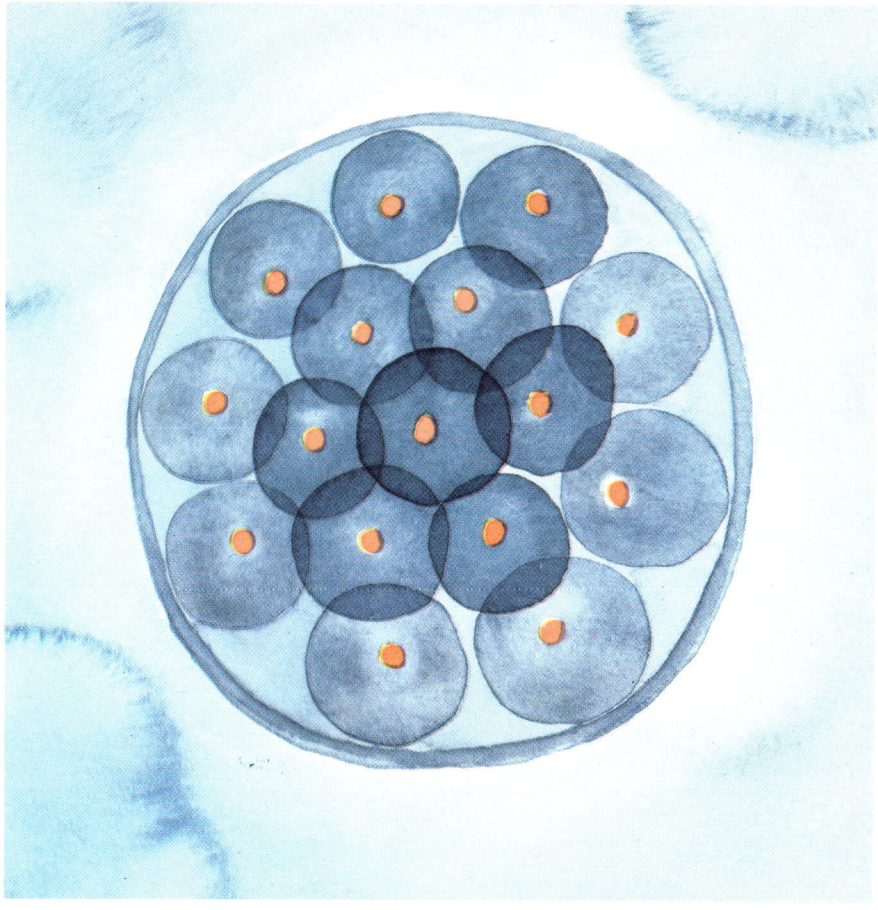

SIXTEEN CELLS

By the fourth day there are sixteen cells. They are still identical to each other though smaller than the first ones, and each contains forty-six chromosomes.

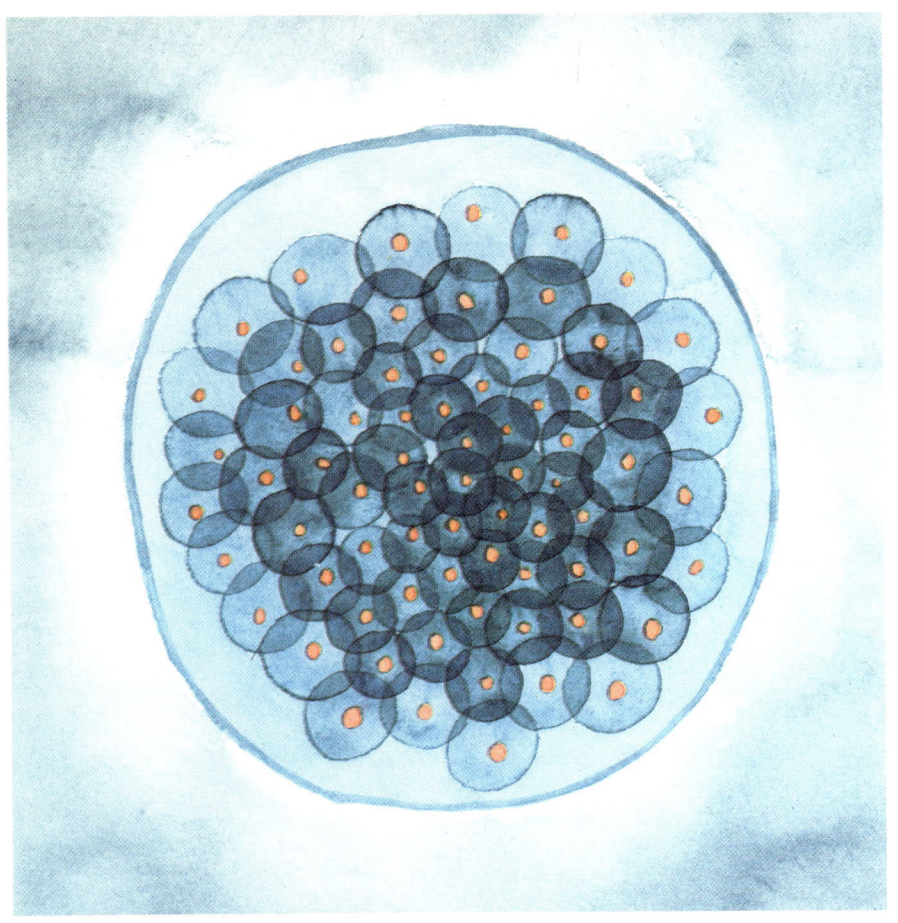

THE MULBERRY

Soon there are thirty-two cells, which continue to divide, doubling their number with each division. The egg has become a compact cluster of cells. The cluster is still round, and looks like a little mulberry.

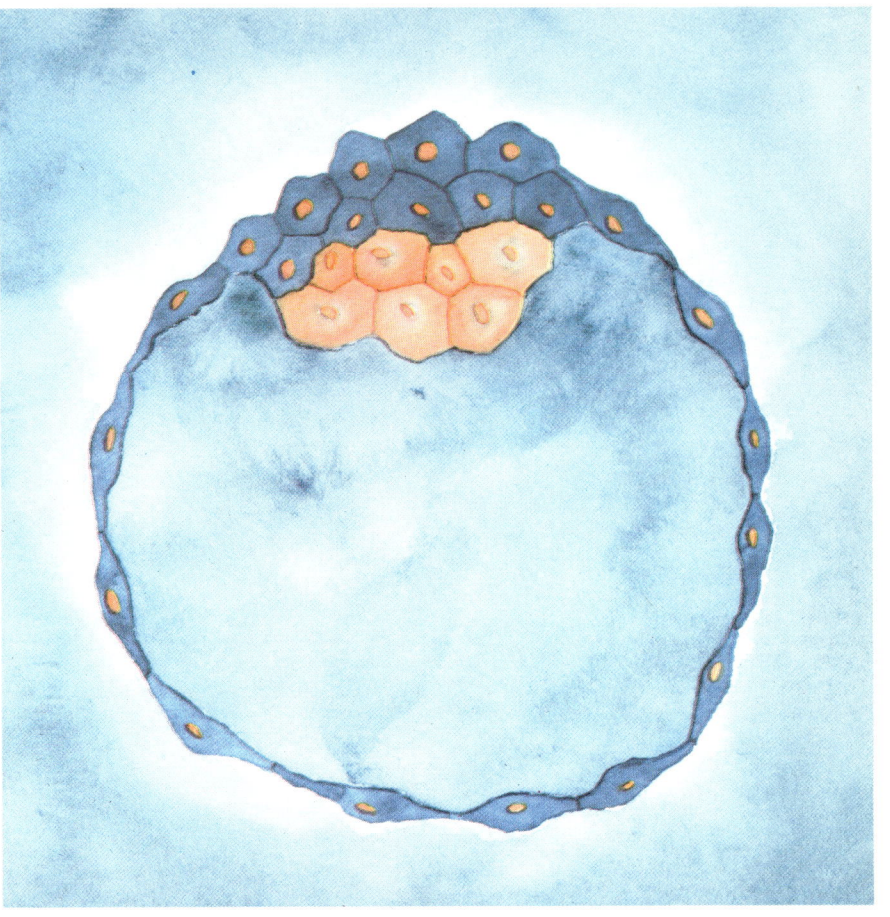

THE EMBRYONIC DISC

The egg itself begins to grow the moment it contains sixty-four cells. On the fifth day, the cluster of cells becomes hollow in the center, forming a space called the amniotic cavity.

The original cells surrounding the cavity will form the placenta, or covering sac, and the umbilical cord for the baby. Within this cavity there is a group of cells in the form of a disc, which will develop into the embryo—the early stage of development when the baby's organs are being formed.

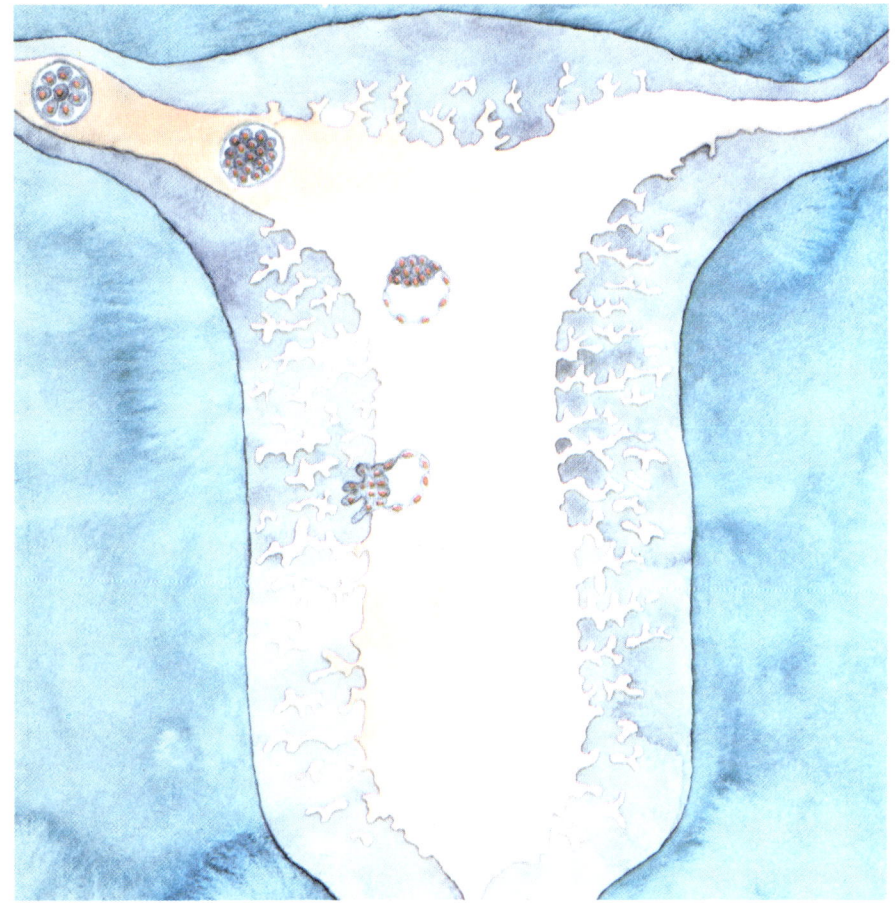

THE EGG REACHES THE UTERUS

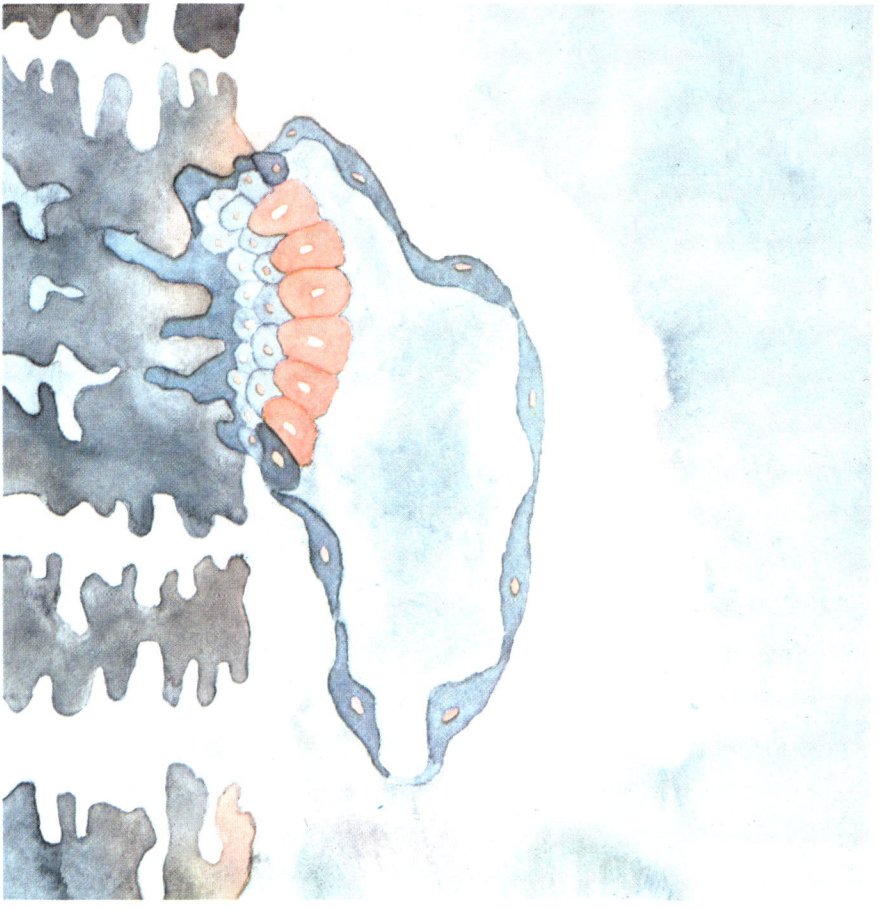

NESTING

The egg has reached the end of its journey down the Fallopian tube. The mucus lining of the uterus is ready to receive it, as plowed land is ready to receive seeds of grain.

   The uterine cavity provides the space in which the egg will live and grow for the next nine months.

The egg is still no bigger than a dot. On the sixth day, it clings to the wall of the uterus, burrows in, and feeds on the liquids it finds there. Tissues heal over it. This is nesting.

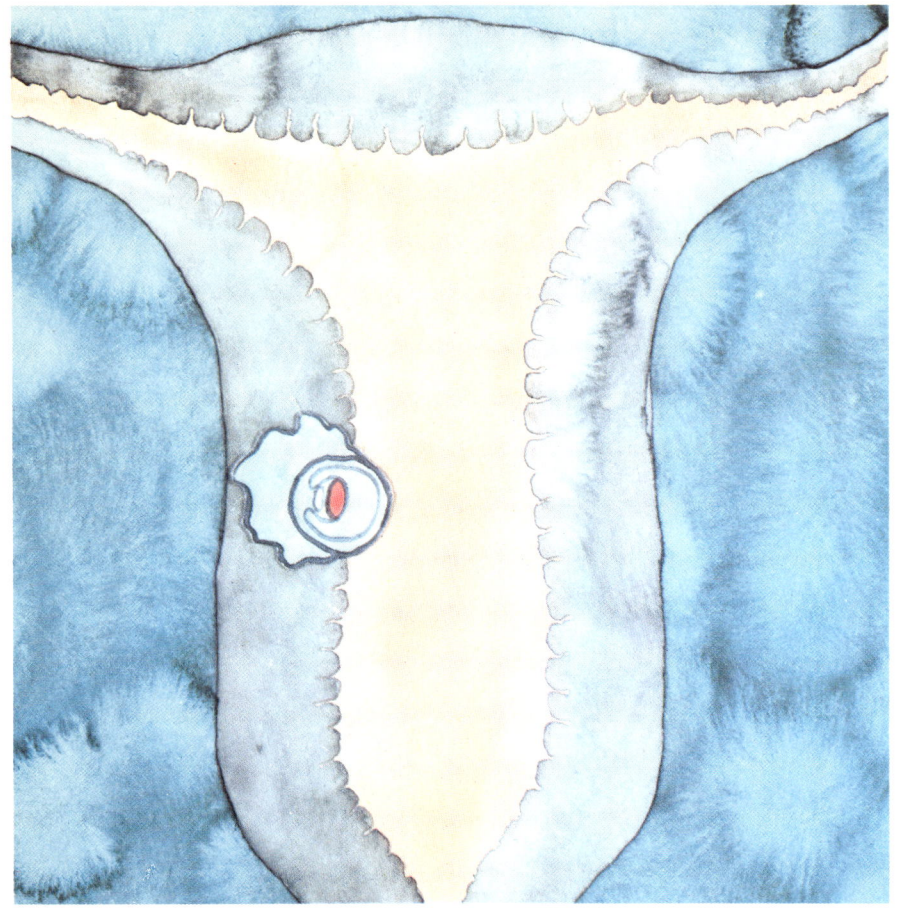

THE EMBRYONIC DISC

When first formed, the embryonic disc is only a tiny fraction of an inch wide, with no distinct shape. Gradually, it begins to take the form of a button or disc. It is this disc which will develop into the embryo. On the fifteenth day a tiny streak, which will later grow into a groove, forms on the back of the disc. This is the beginning of the future spinal cord.

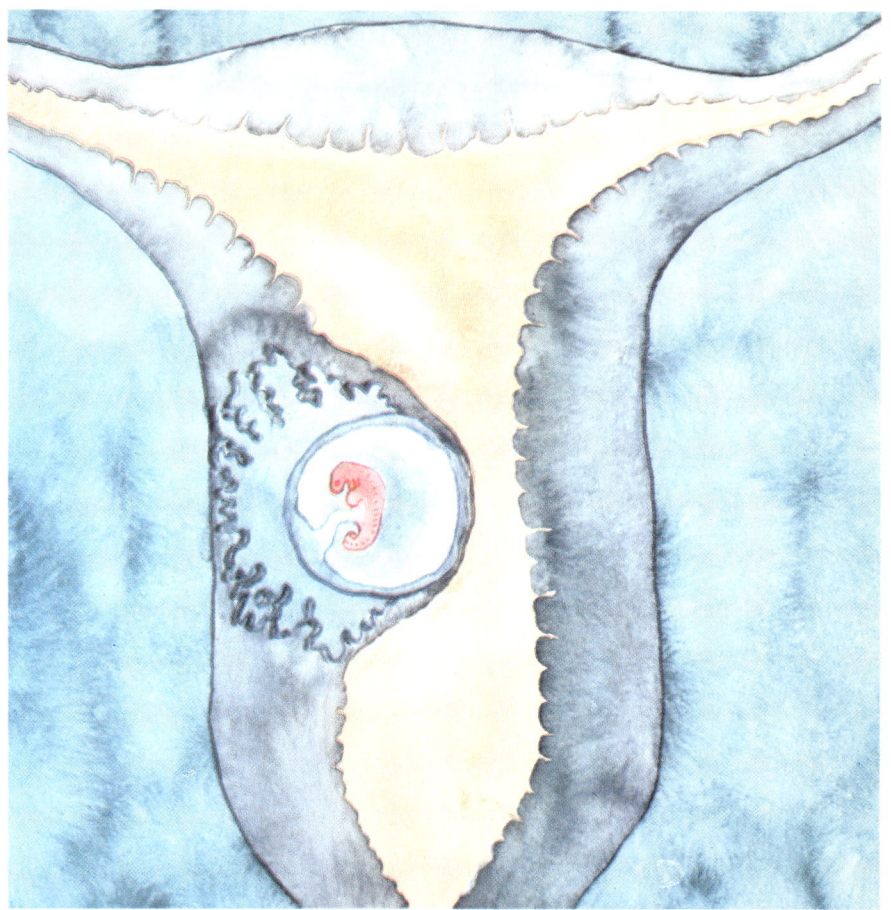

THE EMBRYO

Next, the disc begins to curve inward and grow thicker. The amniotic cavity enlarges and envelops what is now the embryo. This cavity is the future "bag of waters," which will enclose the fully-formed embryo—called a fetus—and protect it against shocks until birth.

In the third week, the first vertebrae (sections of the backbone) appear. A bulge shows the beginning of the head. At the opposite end of the embryo is a tail, though not nearly as pronounced as the head. There is as yet no sign of the arms or legs. The heart starts beating about the twenty-fifth day. The embryo curls up and takes on the shape of a strange little animal.

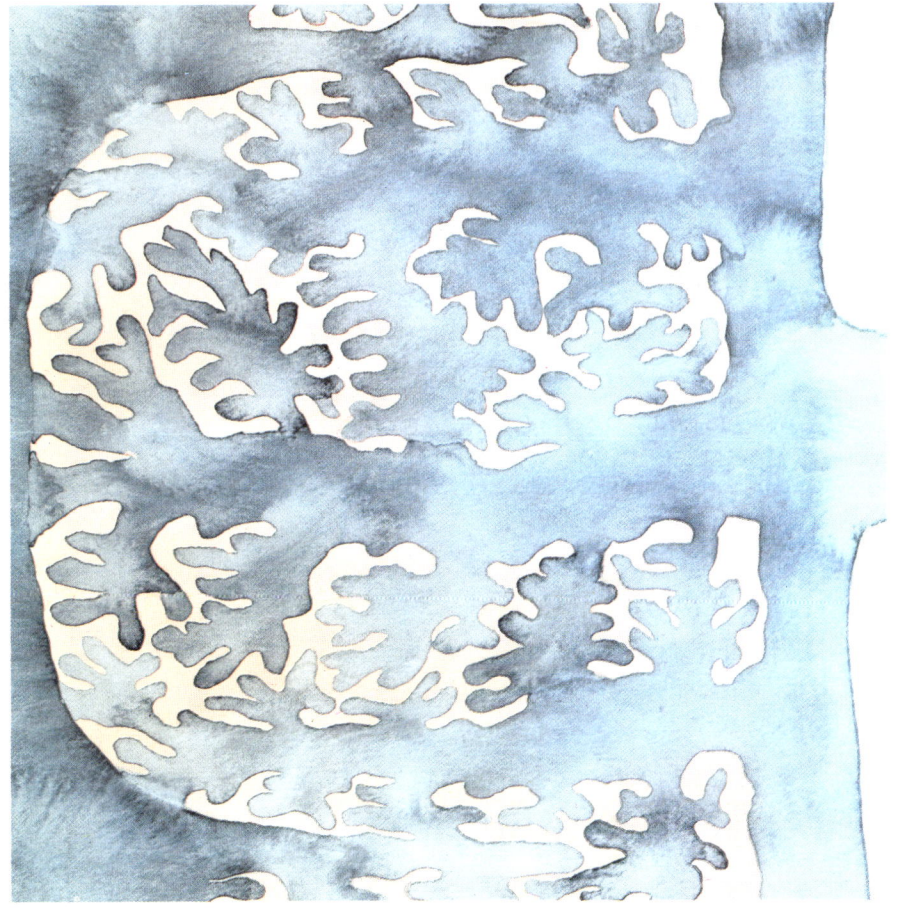

THE PLACENTA, OR TREE OF LIFE

In the first months, the embryo receives its nourishment from the rich blood supply in the lining of the uterus. Later, in the fourth month, the fetus grows rapidly and must have food, oxygen, and water. The mother's body supplies all these things by way of the placenta, the spongy material attached to the uterine wall. Through the placenta the fetus receives life-giving blood, just as a tree receives nourishment from the soil through its roots.

THE UMBILICAL CORD

The umbilical cord is the baby's lifeline. One end of it connects with the placenta, the other leads into the baby's belly at the point which will become the navel. This umbilical cord, a bright, bluish-green tube about twenty inches long, loops around and around in a spiral, and is long enough to allow the fetus to move around freely.

The umbilical cord brings fresh mother's blood from the placenta, and returns used baby's blood. Both kinds of blood flow very fast through the cord—so fast that the cord is always stiff, like a garden hose filled with water.

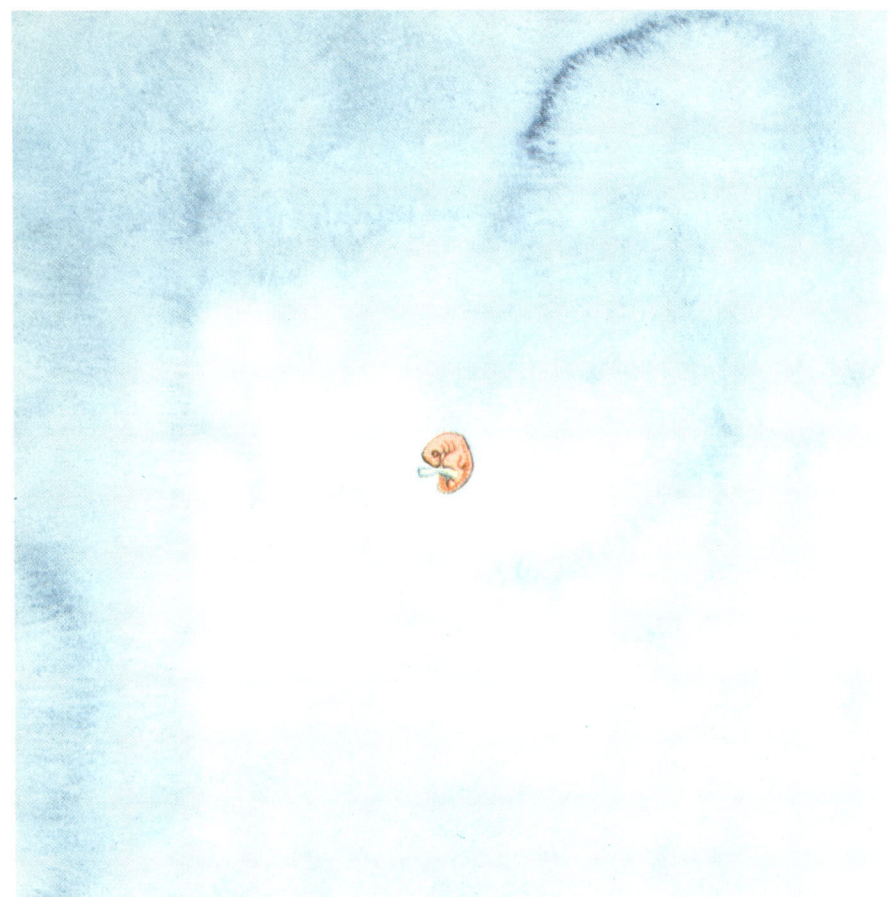

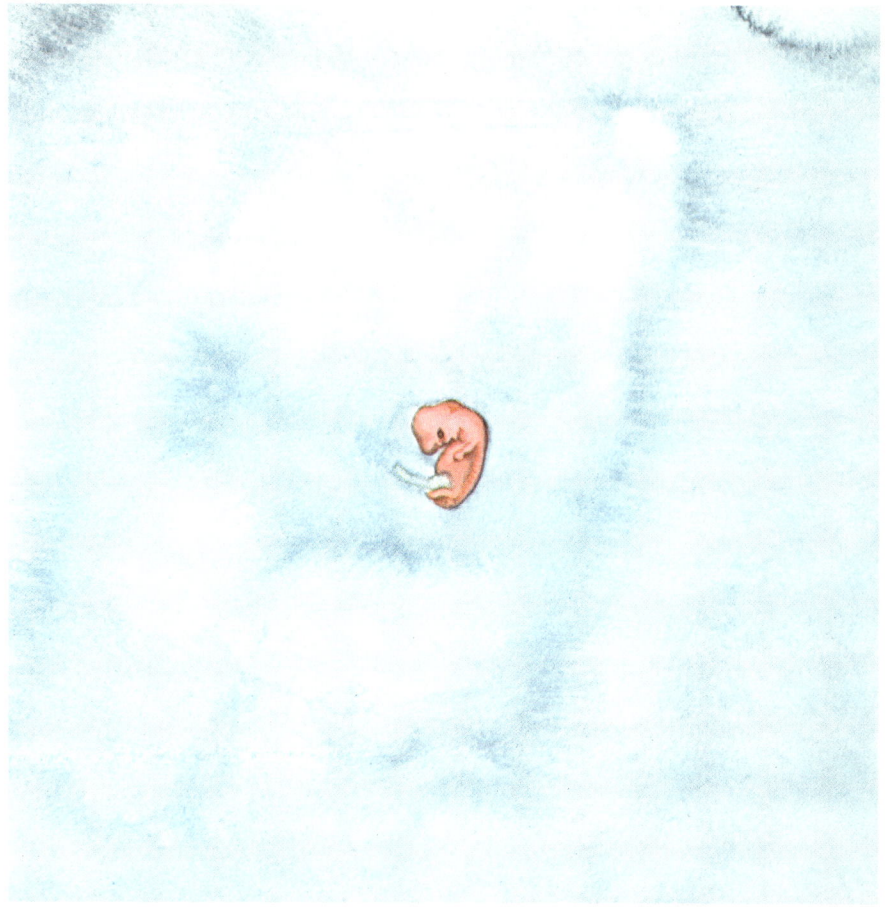

THIRTY-SEVEN DAYS

The embryo is now about one tenth of an inch long. The buds of the arms and legs appear. The head is resting on the chest. It is now possible to make out the head's primitive oral cavity, which will later develop into the nose and mouth. The tail of the embryo begins to shrink and will eventually become the tip of the spine.

FORTY DAYS

The embryo is about one third of an inch long. The future baby looks like a huddled-up reptile. The arm buds have flattened out—a sign that the hands have taken shape. The leg buds appear soon after the hands. The eyes are easy to identify. Two holes show where the ears will be. The formation of the face begins.

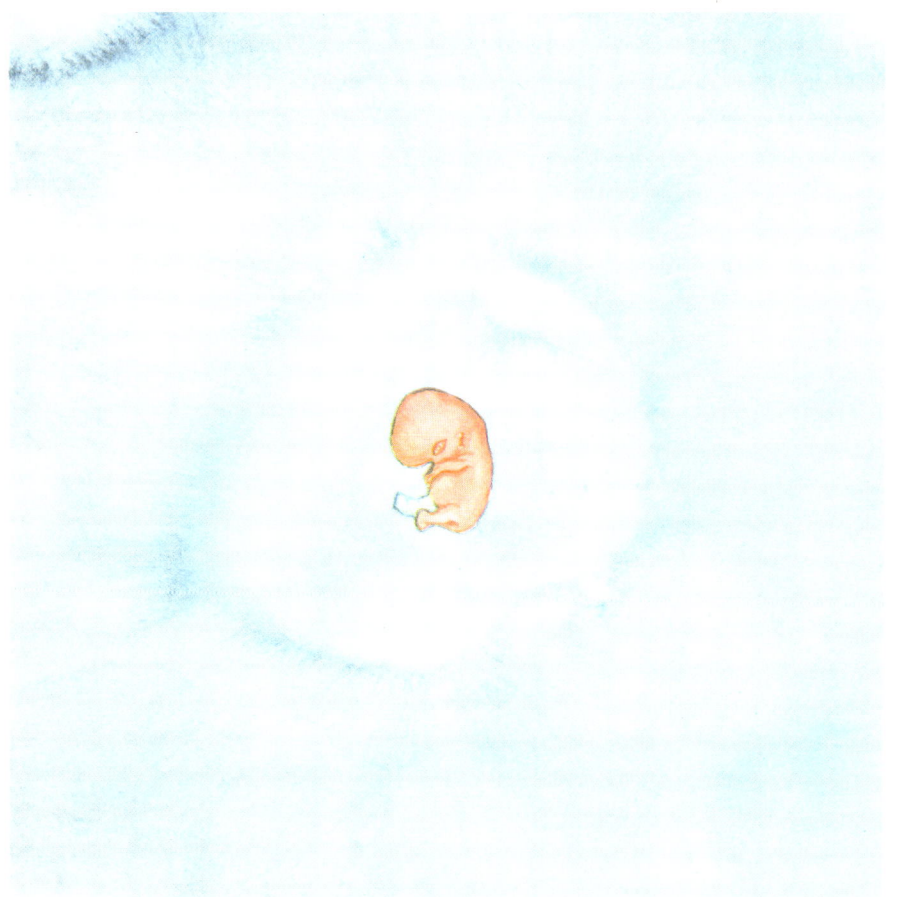

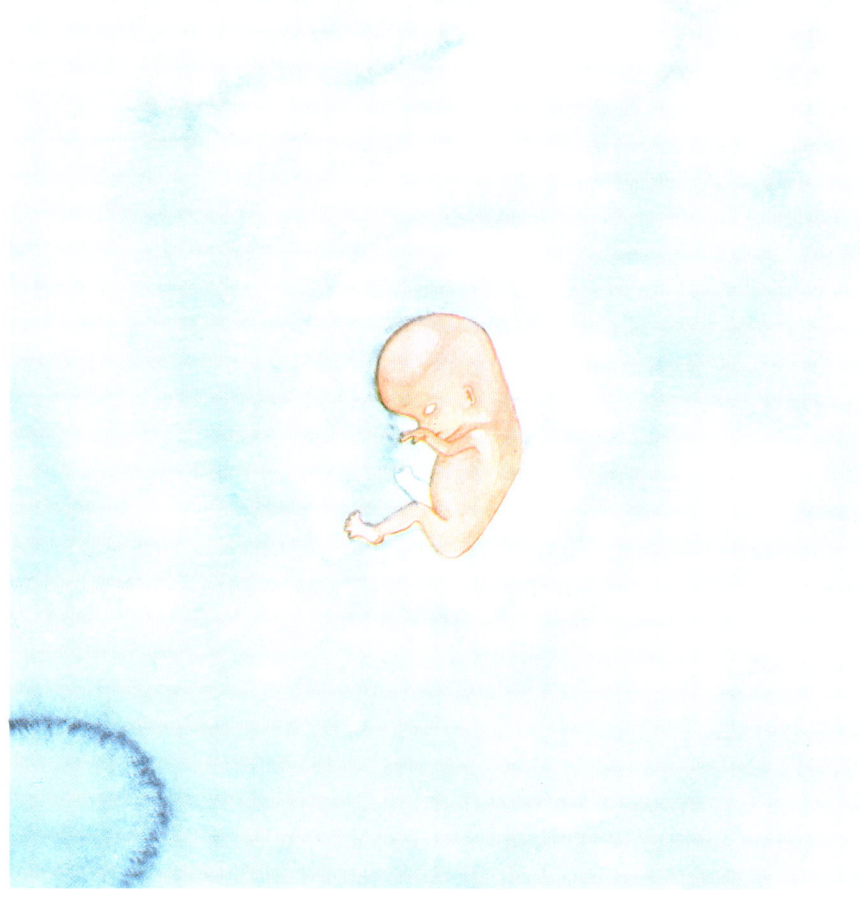

FORTY-SIX DAYS

Now the embryo is about three quarters of an inch long. The arm and leg stems are longer.
The knee and elbow joints can be seen. The milk teeth are forming in the gum ridges. The embryo makes feeble movements.

SIXTY DAYS: THE FETUS

After the seventh week the embryo is called a fetus. At this time, it is one inch long, and has a distinctive human form.

THE HAND: THIRTIETH DAY

At first, the baby's hands are like flower buds. By the thirtieth day, the arms are little stems folded along each side of the body.

THE HAND: THIRTY-THIRD DAY

By the thirty-third day, the shape of the fingers can be seen at the end of the hand. The feet appear almost at the same time.

THE HAND: SEVENTH WEEK

In the seventh week, the thumb and fingerprints can be seen. The ends of the fingers are flat, like frogs' feet.

THE HAND: THIRD MONTH

In the third month, the flat-ended pads of the fingers grow smaller, then disappear; the hand looks like a true human hand. The feet follow the same development as the hands, but more slowly.

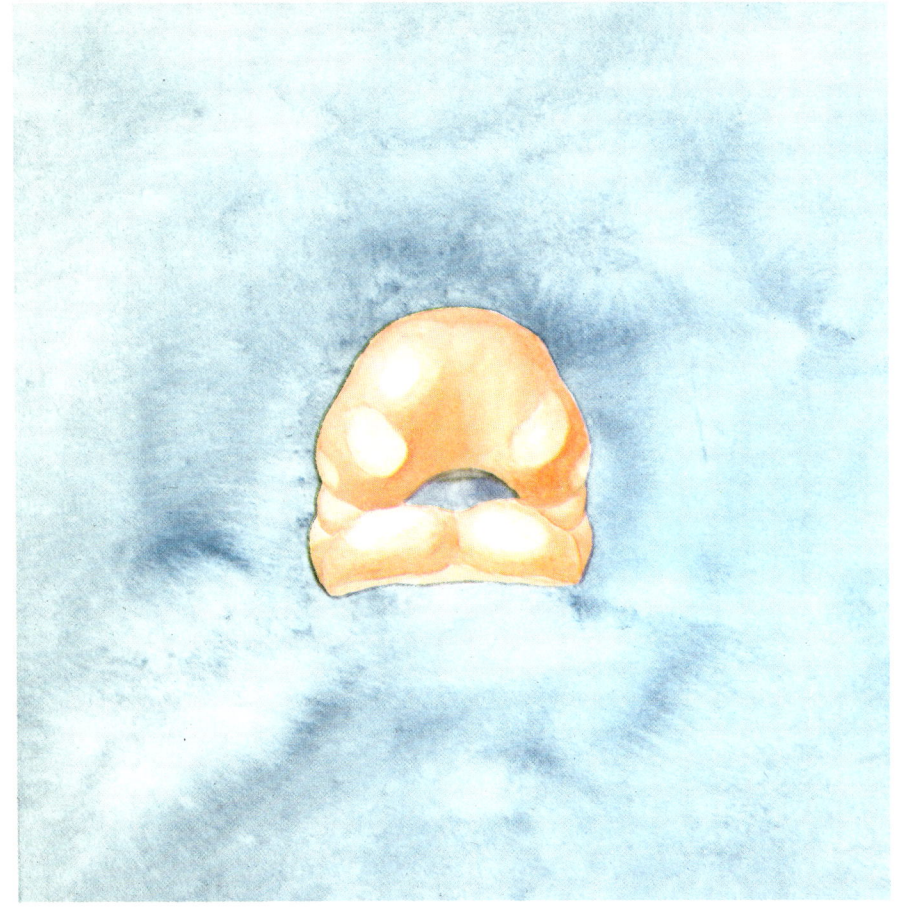

THE HEAD

When the baby's head first begins to form, it looks like the head of a tadpole.

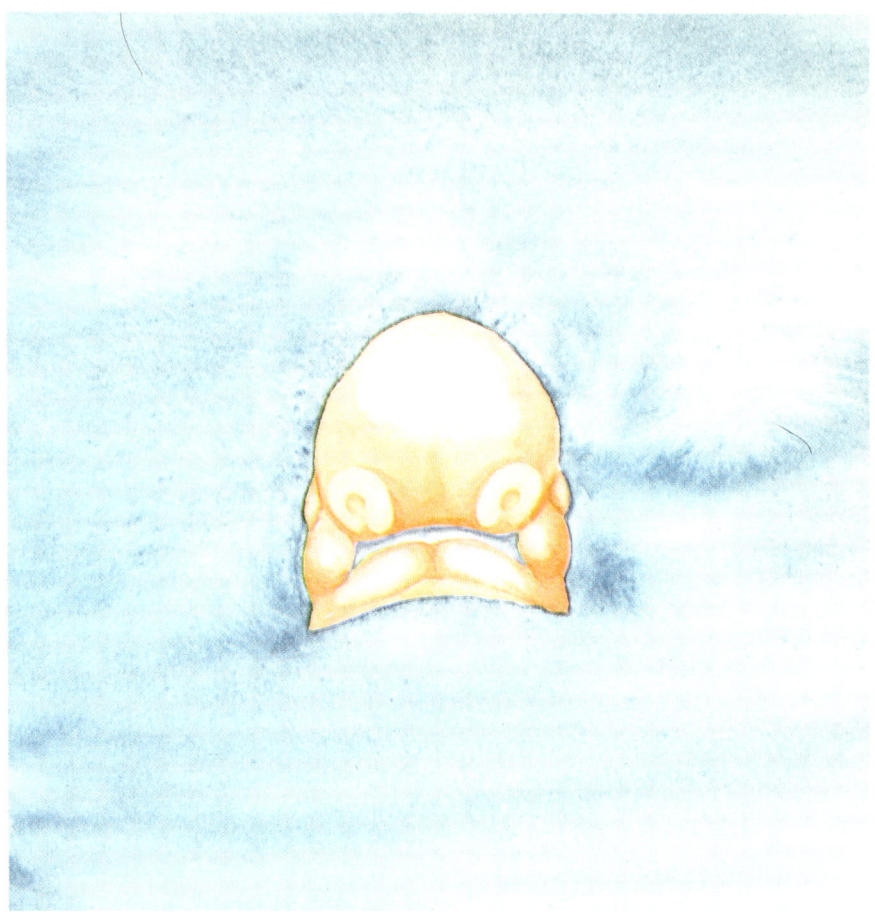

THE HEAD: SIXTH WEEK

By the sixth week, the eyes, which are easily made out, are beginning to show their color. The formation of the face continues. The nostrils are outlined—very wide apart. The mouth is still nothing more than a broad gash, like the mouth of a fish. The brain is very large in relation to the rest of the body.

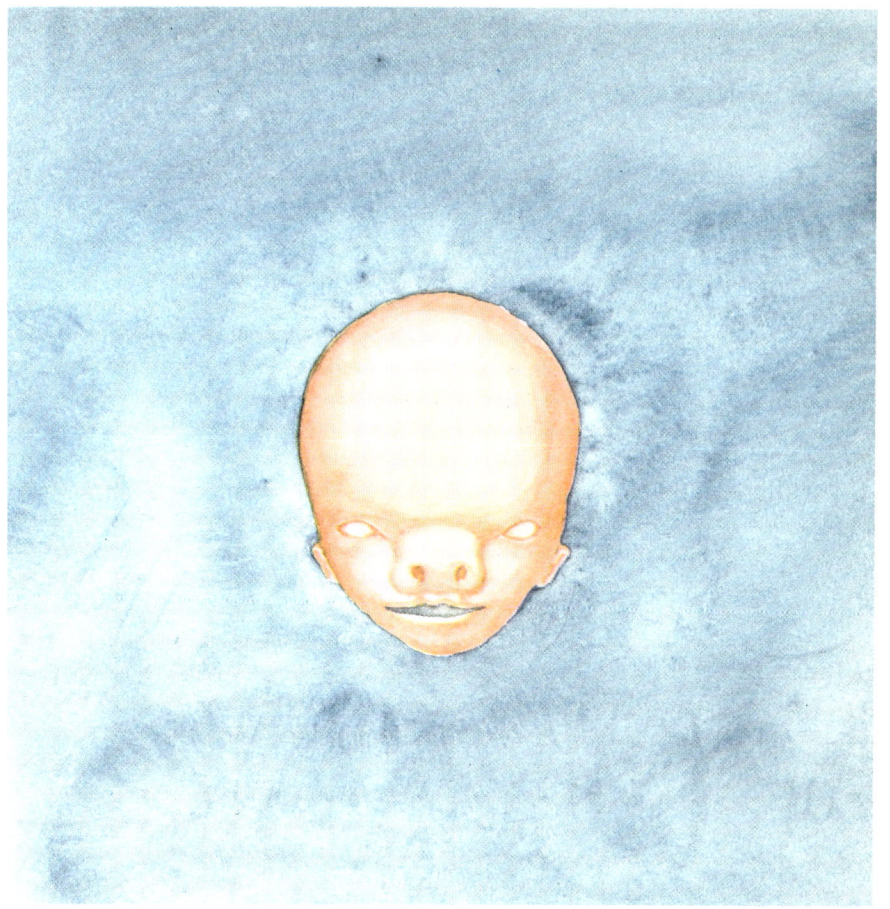

THE HEAD: SEVENTH WEEK

THE HEAD: TENTH WEEK

By the seventh week, the neck of the fetus has begun to take shape. The eyes, lidless like those of a bird, are located on either side of the head. The nose is flat, but the nostrils have moved closer together.
The mouth is still very broad, extending across the lower part of the face. The ears are still very small, and located far down on the sides of the head. By now, the face looks like that of a wrinkled, little old person.

By the tenth week, the fetus has a truly human face: the eyes are now in place; the eyelids have developed; the lower lids close against the upper, keeping the eyes sealed until the sixth month.

  Twenty teeth have taken root, although they will not start forming as real teeth until the eighteenth week. The lips are outlined, and the tongue has formed. Fine, downy hair appears on the face and the body. This is the lanuga, or hair coat, which will disappear following birth.

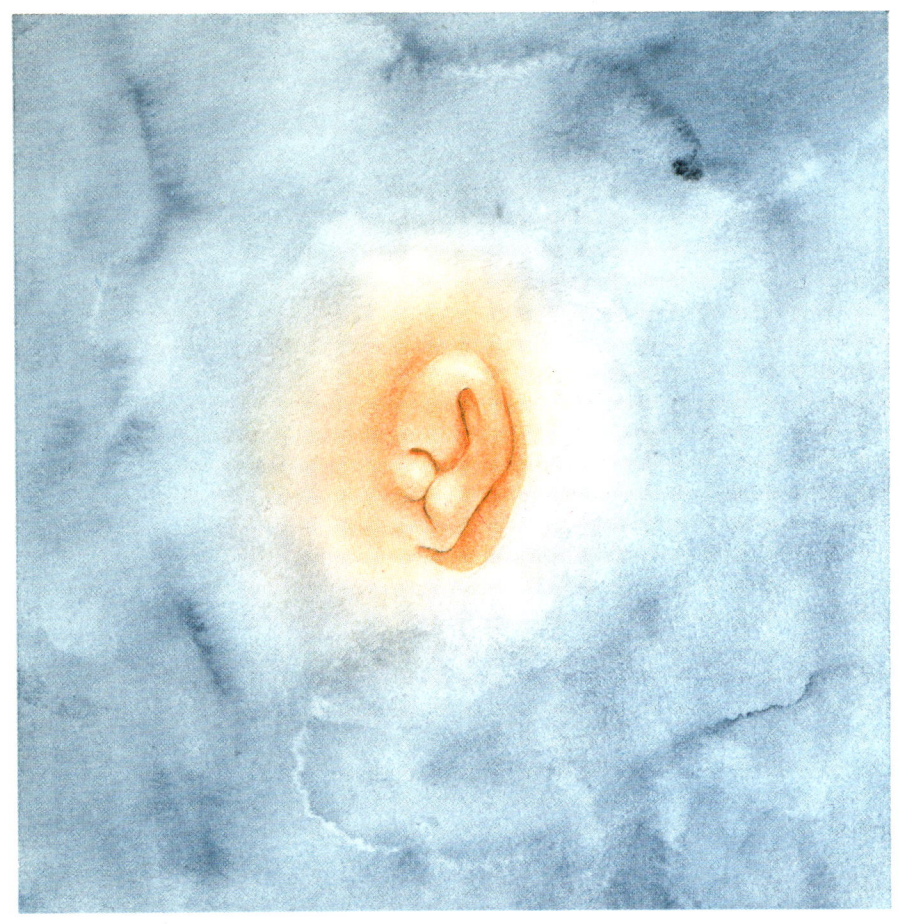

THE EAR: SECOND MONTH

During the fifth and sixth weeks, the ears develop from two folds of tissue along the sides of the head. They are shaped like question marks. And they can already hear the sounds of their inside world: something like the rushing of a waterfall, and a kind of jangling.

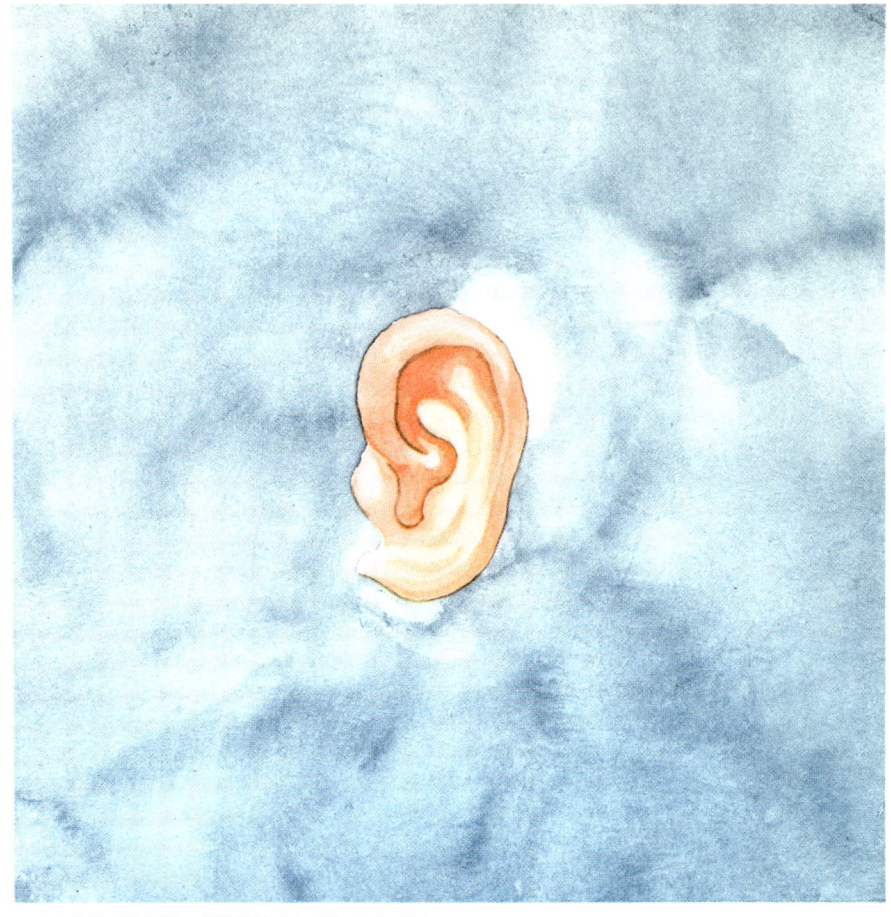

THE EAR: NINTH MONTH

By this time, the fetus, with its fully-formed ears, is already getting to know the noises of the outside world where it will soon begin its life as a human being.

SEX: GIRL

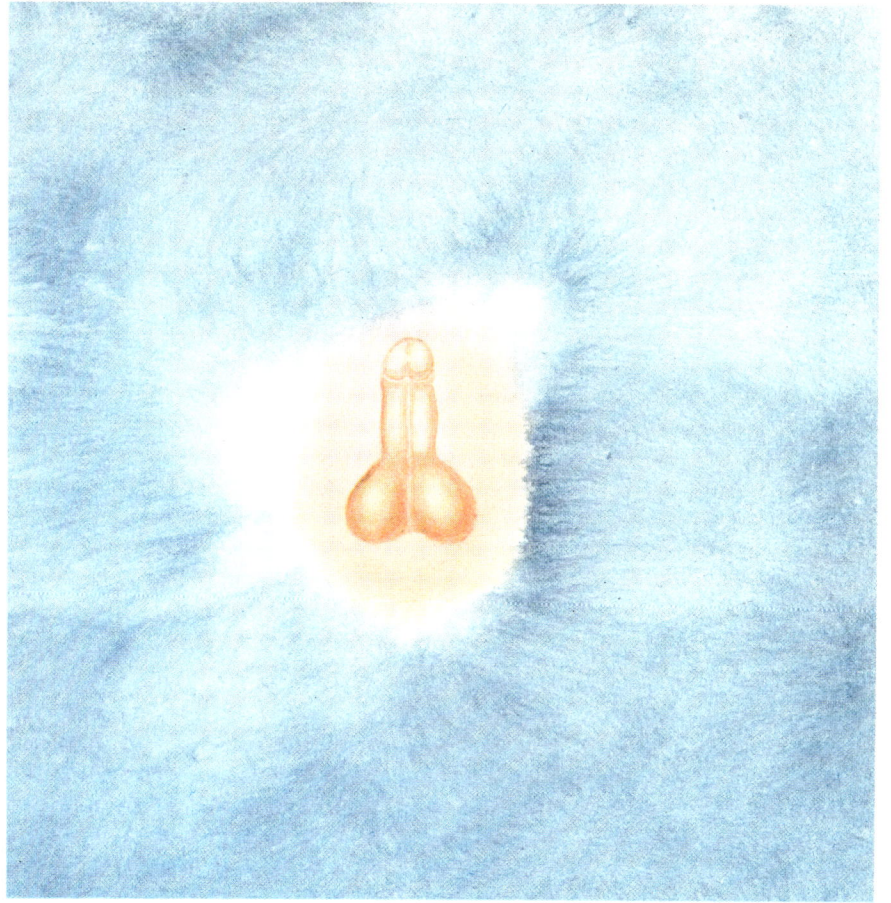

SEX: BOY

By the end of the third month, the fetus is recognizably either a girl or a boy. Its sex was decided at the very moment of fertilization. Of the twenty-three chromosomes in both the sperm and the ovum, twenty-two determine the future baby's personality, features, and the rest of its physical and mental makeup. The twenty-third chromosome determines its sex.

In the ovum, this sex chromosome is always of the X, or female, type. In the sperm, it may be either an X or a Y (male). The baby's sex depends upon whether this twenty-third chromosome in the sperm is an X or a Y. If it is an X, the baby will be a girl.

If the twenty-third chromosome in the sperm is a Y, the baby will be a boy.

The other twenty-two chromosomes are often called "the craftsmen of heredity," since they pass on physical features and personality traits from generation to generation. But it is important to remember that each parent passes on his or her traits to only one half of the child. The child, uniting characteristics from two different people, is itself a unique individual.

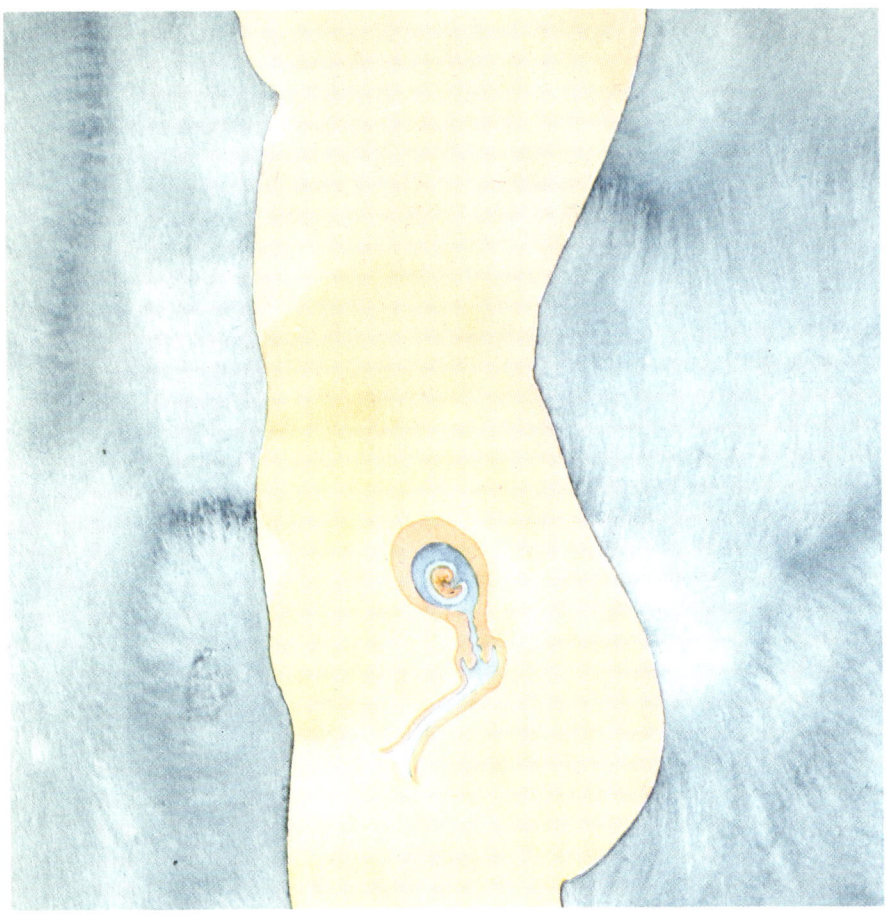

EARLY STAGES OF GROWTH

From the very first days in the mother's womb, the baby grows at a dizzying rate. The embryo grows faster than the fetus, the fetus faster than the baby. The baby ages faster than the adult, and the adult faster than the old person. Curiously, it is the old person who ages most slowly.

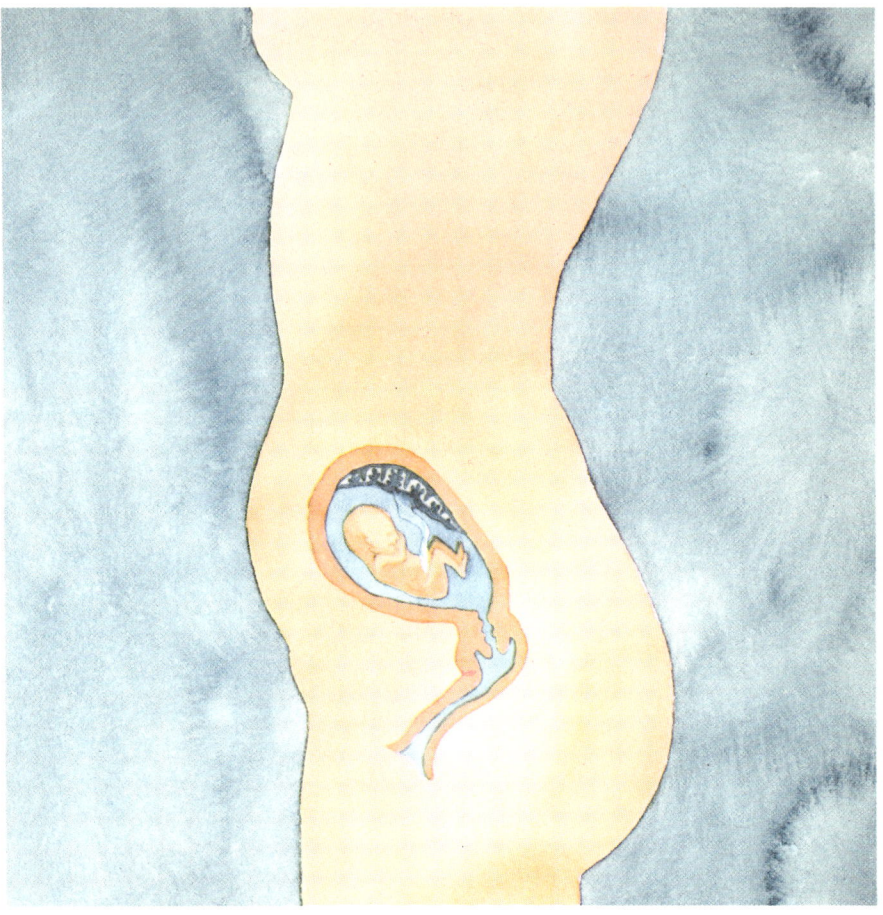

FOURTH MONTH: QUICKENING

Toward the end of the fourth month, the baby weighs over one pound and is about ten inches long. This is the month when the mother feels something like the fluttering of wings inside her body. This movement is called quickening. It soon changes into real thumps. The baby sleeps and wakens. And kicks! It moves about freely in a watery world, seeking the best position or "lie."

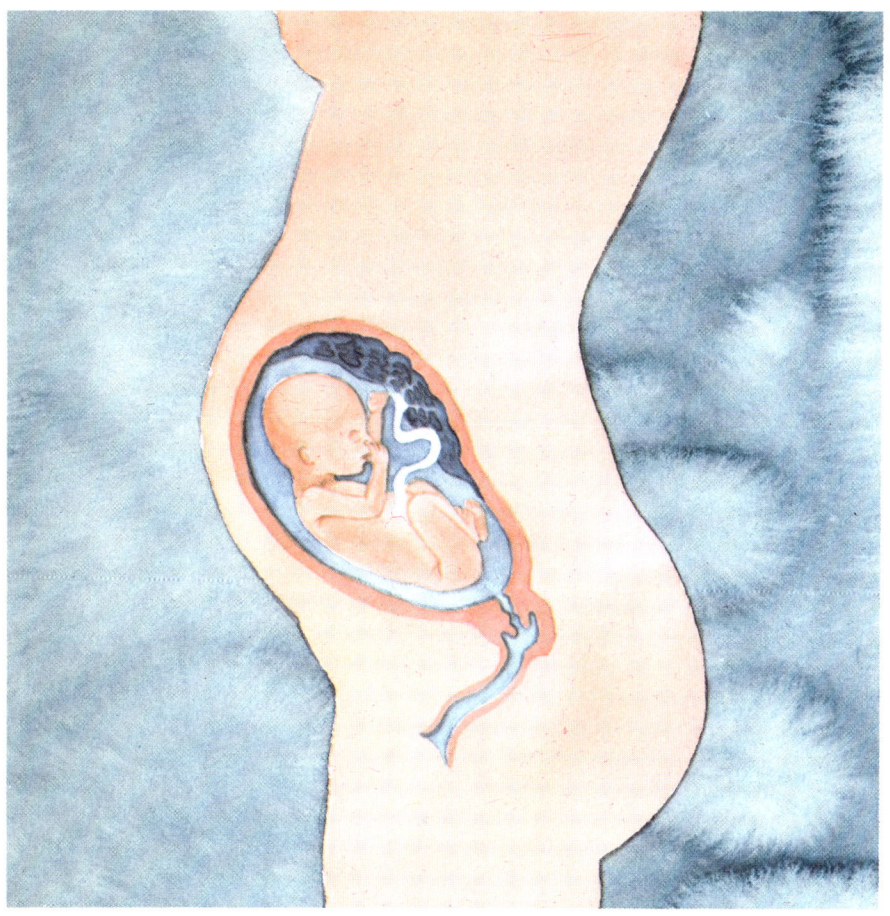

SEVENTH MONTH

In the course of the seventh month, the baby's weight increases to almost three pounds. The water around it protects it from jolts from the outside, and keeps it at an even temperature. The baby sucks its thumb, opens its eyes and then closes them again.

But once into the eighth month, the baby is too big to move around very much. Then it settles into its favorite position.

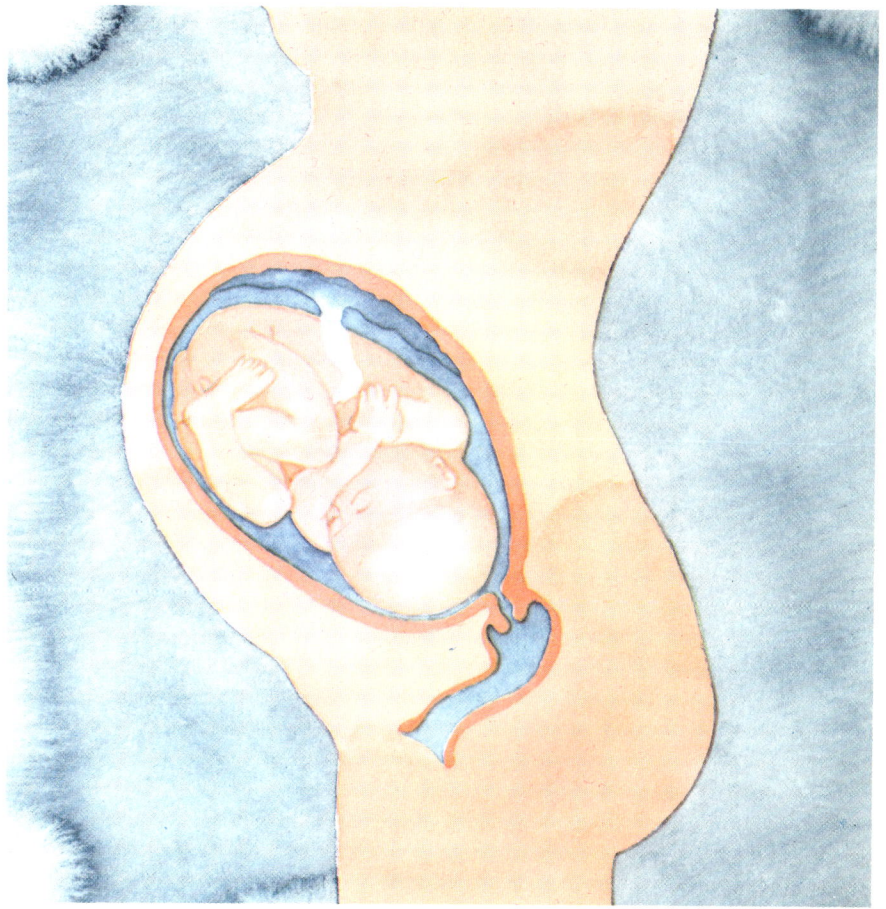

NINTH MONTH

By the end of the ninth month, the average baby weighs a little over seven pounds and is about twenty inches long. It is ready to leave its mother's womb, and is taking up all the space available. All curled up, with its head in a downward position, its knees against its nose, its thighs tight up against its torso, and its arms crossed, the baby is ready for its journey outward. It now presses its head against its mother's cervix—the narrow end of the birth canal leading into the vagina.

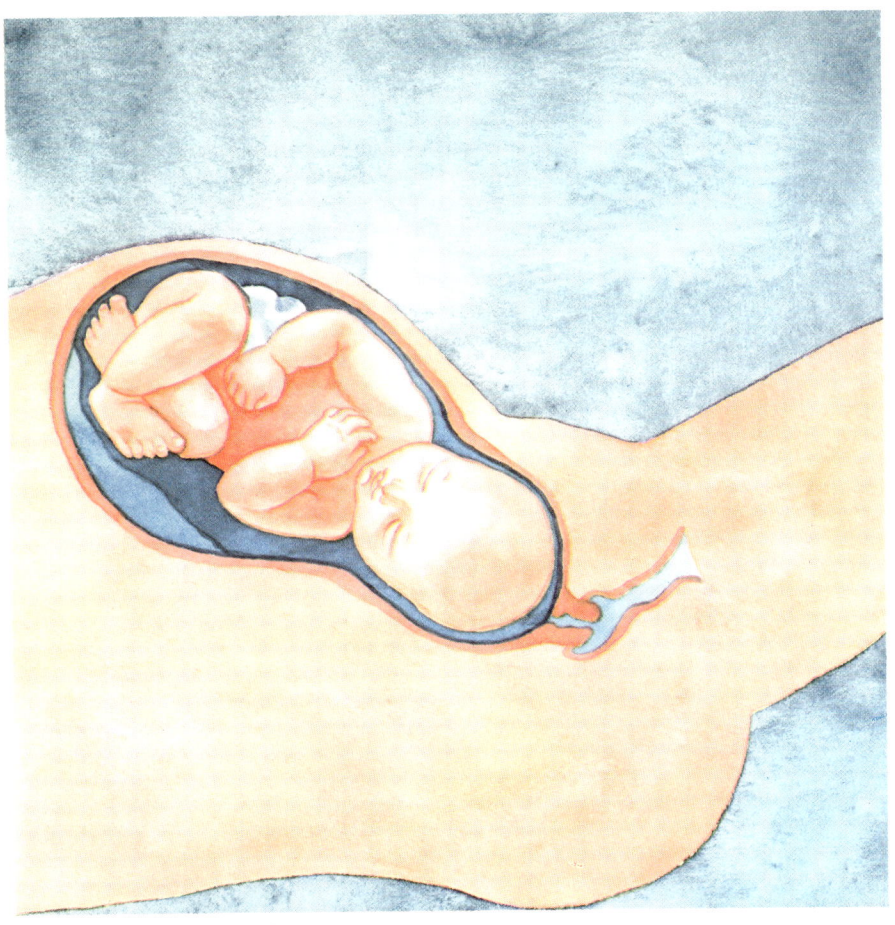 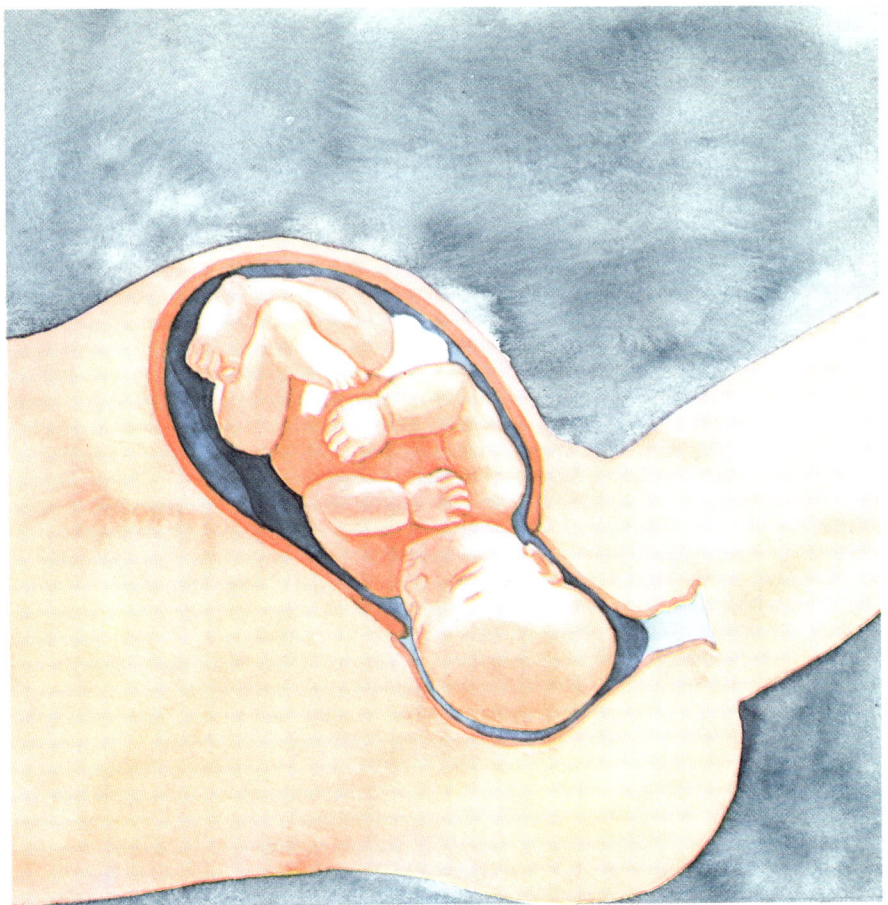

FIRST STAGE OF LABOR

The baby is ready to be born. Its head is pressed hard against the cervix. The cervix will soon dilate, or spread open, and the baby's head will fit itself to the size of the birth canal.

Now come the contractions—the squeezing by the muscles of the uterus. This is the first stage of labor. Under this pressure, the bag of waters usually breaks.

The contractions become more frequent. Each one helps to widen the birth canal and the head drops lower and lower.

SECOND STAGE OF LABOR

As the baby moves farther down, it shifts its position so that the narrowest part of the head enters the birth canal. The contractions become stronger and more frequent. When the canal is completely open, the second phase of labor begins.

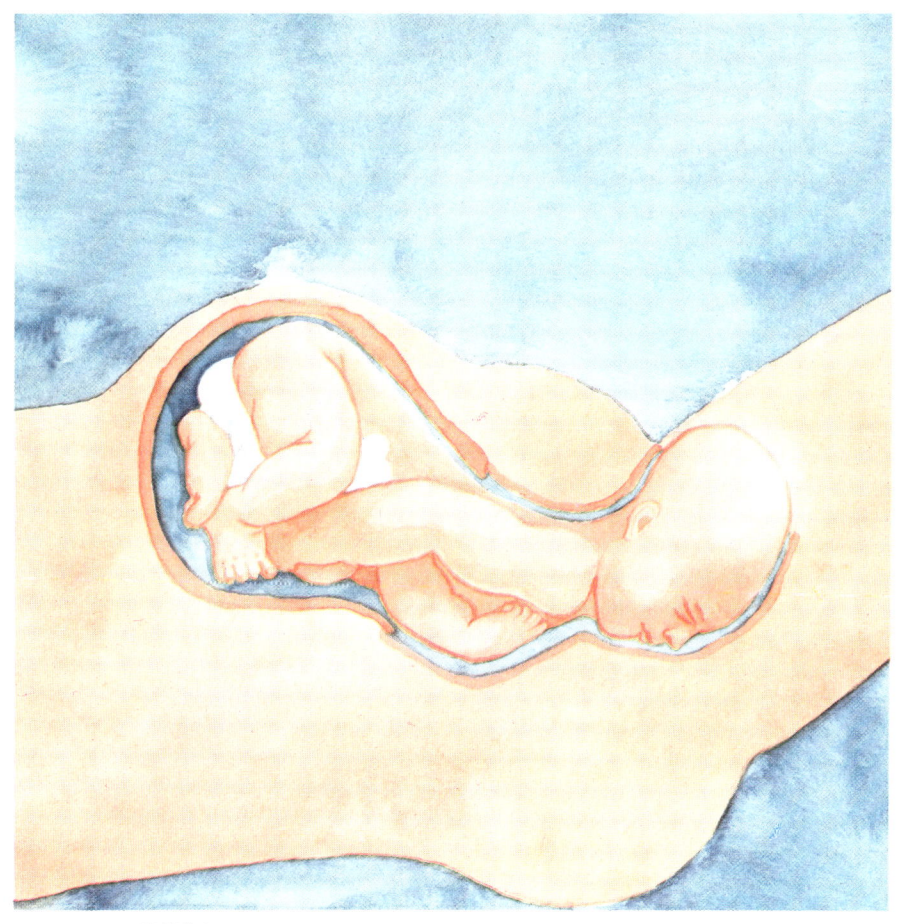

CROWNING

The contractions now reach their strongest peak. Powerful uterine muscles press down against the baby, pushing it through the vagina.

Under this great pressure, the baby's head is pushed out through the vagina into the world. This is called crowning.

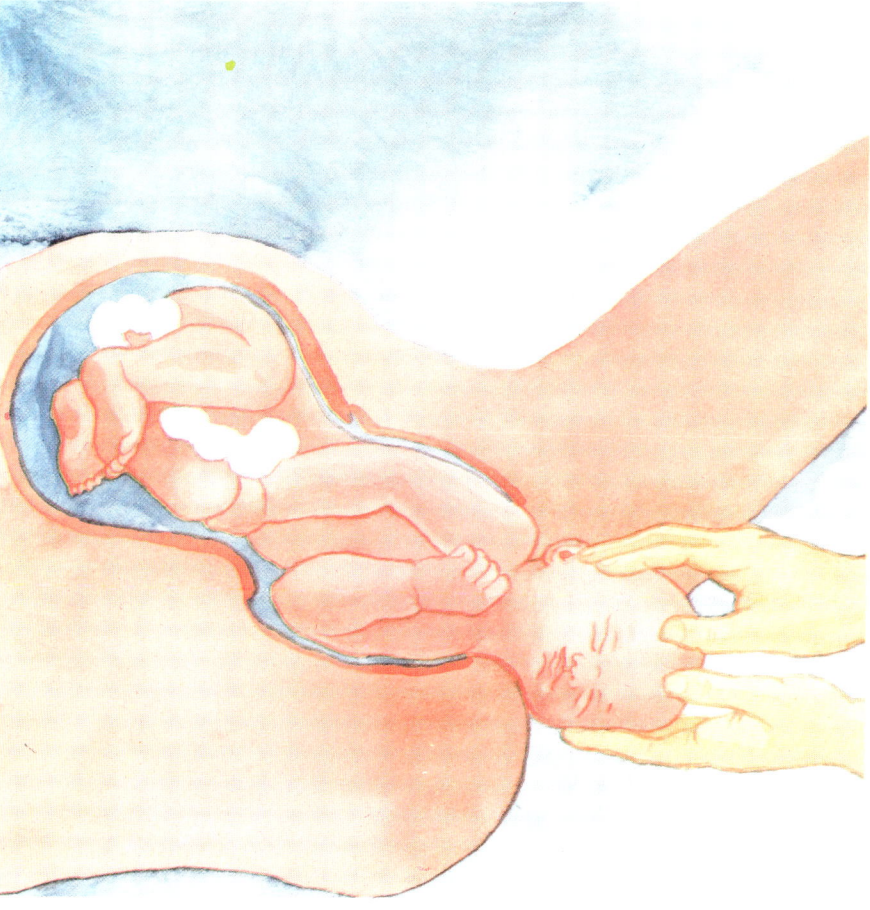

EXPULSION

When the head is completely out, the baby turns it to get it in line with its shoulders. Then it turns its whole self, as the doctor guides the shoulders and the rest of the body out.

THE CHILD IS BORN

The child is born. It is gasping for air. The chest expands as the lungs fill with air for the first time. The vocal chords vibrate. And the child utters its first cry.